P.G. Morgan, J.A. Bartrum

List of the Minerals of New Zealand

Salzwasser

P.G. Morgan, J.A. Bartrum

List of the Minerals of New Zealand

1. Auflage | ISBN: 978-3-84604-683-8

Erscheinungsort: Frankfurt, Deutschland

Erscheinungsjahr: 2020

Salzwasser Verlag GmbH

Reprint of the original, first published in 1913.

NEW ZEALAND

Department of Mines.

Geological Survey Branch (P. G. Morgan, Director).

LIST

OF THE

MINERALS OF NEW ZEALAND,

BY

P. G. MORGAN AND J. A. BARTRUM.

Issued under the authority of the Hon. WILLIAM FRASER, Minister of Mines.

WELLINGTON, N.Z.
BY AUTHORITY: JOHN MACKAY, GOVERNMENT PRINTER.
1913.

LIST

OF THE

MINERALS OF NEW ZEALAND.

INTRODUCTION.

For many years there has been an urgent need for the compilation of a complete list of mineral occurrences in New Zealand. Such a list is required by the geologist, the prospector, the miner, and all interested in the development of the mineral resources of this country. Its preparation has been contemplated for some years, and in consequence a considerable amount of material has been collected by members of the Geological Survey Staff. It was not intended to publish this matter until a complete handbook to the minerals of this country could be written, but a request for a list of New Zealand minerals to accompany the "Catalogue and Description of Exhibits of the Mines Department" prepared for the Auckland Exhibition has led to the compilation of the present list. Owing to the available time being extremely short, it has not been possible to include in the list all the data that could be obtained. Mainly on this account much interesting information contained in the accessible literature, which would be of value to prospectors and others, has been omitted. For a similar reason, many references to localities where useful minerals occur, but as a rule in small quantity only, are not here given. Perhaps the most important of such omissions are data relating to the clays, limestones, and mineral waters of this country, as well as to ores of economic importance. Much of this material, it is true, does not strictly belong to the domain of mineralogy, but is certainly not out of place in a publication intended for the benefit of the general public.

In referring to authors, the present writers, owing to want of time, have in many cases given only the first reference found, and not the original source, which is oftentimes difficult, or even impossible, to trace. A few very doubtful occurrences have been purposely omitted, but others which are probable rather than certain are included, and on the whole the list is believed to be tolerably complete as regards positively identified mineral species. For the sake of rendering the list as useful as its limits permit, the writers have followed the plan usually adopted in similar publications—namely, of including various substances of economic value, which in strictness are not minerals.

An incomplete list of the literature consulted is given below, but it may be observed that the chief single source of information has been the Laboratory Reports of the late Mr. W. Skey and of Dr. J. S. Maclaurin. The chemical data in these reports are of a high order of reliability, but have necessarily been omitted from the present publication. Since in most cases the analyst had to accept as correct the information supplied by the senders of mineral specimens, the localities are often vaguely recorded, and in a few cases are no doubt wrongly stated.

The greater part of the work connected with the writing of this catalogue has been done by Mr. J. A. Bartrum, who has indexed the numerous mineral occurrences recorded in the annual reports of the Colonial or Dominion Analyst, and has brought most of the other material into proper shape for publication.

LITERATURE.

The principal publications consulted in connection with the preparation of the list are,—

1878. Liversidge, A.: "Notes on Some of the New Zealand Minerals belonging to the Otago Museum, Dunedin." Trans. N.Z. Inst., vol. x, pp. 490–505.

1882, 1883. Cox, S. H.: "Notes on the Mineralogy of New Zealand." Trans. N.Z. Inst., vol. xiv, pp. 418–450, and vol. xv, pp. 361–408.

1892. Hector, James: "Minerals of New Zealand." Rep. of Geol. Expl. during 1890–91, pp. 105–120. Revised from Rept. Aust. Assoc. Adv. Sci., vol. ii, 1889, p. 269; and reprinted in the New Zealand Mines Record, vol. 3, 1899–1900, pp. 256–263, and in the New Zealand Mining Handbook, 1906, pp. 500–519.

1909. Marshall, P.: "Additions to the List of New Zealand Minerals." Trans. N.Z. Inst., vol. xli, pp. 105–110.

1910. Park, James: "The Geology of New Zealand" (see List of Minerals, pp. 398–408).

1867–1901. Skey, W.: Annual Reports of the Colonial Museum and Laboratory, Annual Reports of the Colonial Laboratory, Nos. 2–34.

1902–1913. Maclaurin, J. S.: Annual Reports of the Colonial Laboratory, Annual Reports of the Dominion Laboratory, Nos. 35–46.

1906–1912. Geological Survey of New Zealand: Bulletins Nos. 1–15.

In addition to the above, a large number of other publications have been consulted, abbreviated references to most of which will be found in the list.

ABBREVIATIONS USED.

The following abbreviations are used :—

Trans : Transactions of the New Zealand Institute. A number indicates the volume.

Lab. : Colonial Museum and Laboratory, or Colonial Laboratory, or Dominion Laboratory. Followed by a number (thus—Lab. 3) indicates the number of the annual report of the Laboratory.

Bull. : Bulletin of the New Zealand Geological Survey (New Series).

G.S. Ann. Rep. : Annual Report of the New Zealand Geological Survey.

G.S. : Followed by a date refers to determinations, as yet unpublished, made by members of the staff of the Geological Survey.

Rep. G.S.: Indicates "Reports of Geological Explorations" published by the earlier Geological Survey. The date refers to the year of publication.
(*R*)*:* Indicates that the occurrence is unconfirmed.
Mines Reports: "Papers and Reports relating to Minerals and Mining"—the annual report of the New Zealand Mines Department.
Mines Record: New Zealand Mines Record.

The name of an author enclosed in brackets with or without a date means that a reference to the mineral has been made by that author. In a very few cases the reference is to a verbal statement. If two names separated by a dash are so enclosed the meaning is that the first author is quoted by the second, but the reference has not been verified.

LIST OF MINERALS.

Actinolite (*var. of Amphibole*): Milford Sound (Hector); Parapara (Cox); Westport (G.S. 1913); various localities in North Westland (Bull. 1 and 6); Ophir (Bull. 5); Paringa River (Cox); Dusky Sound (Park); Awatere Valley (Buchanan, Rep. G.S. 1868). It occurs usually in schists and hornfels.

Adularia: See Valencianite.

Ægirine (Ægirite—*var. of Pyroxene*): Dunedin, in alkaline volcanic rocks (Marshall and others).

Ænigmatite (Cossyrite—*var of Amphibole*): In volcanic rocks at Dunedin (Marshall); Campbell Island (Marshall, Trans. 41).

Agate: See under Quartz.

Albite (*Soda-feldspar*): Occurs as a constituent of various igneous rocks. Reported from Maori Point, West Coast; Wilkin River; Makarora; Dun Mountain; George Sound (Hector, Haast, and Davis); also many other localities.

Algodonite (*Copper Arsenide*): Whangapirau (Skey, Lab. 31).

Allanite (Orthite—*var. of Epidote*): Westport district, in diorite (G.S. 1913); (?) Parapara (Thomson, Trans. 42).

Almandine (*var. of Garnet*): Malvern Hills (Haast); Otago (Marshall, Trans. 41).

Alum: Pomahaka, as a product of pyritous shale (Hector 1862); Puai Island, Waikouaiti (Hochstetter 1860); Tokomairiro (Hector 1862); D'Urville Island (Hackett 1886); Kaikorai, Dunedin (Skey, Lab. 28); Mokihinui (Handbook of N.Z. Mines 1887, and G.S. 1913); abundant at Waiotapu and Orakei Korako (Marshall 1912); Pukeiwitahi, Shag Point (*R*); Coromandel (Skey, Lab. 19).

Alunite: Ohinemutu (Skey, Lab. 24) and Rotorua (Ulrich), deposited by geysers; Patua Ranges (Skey, Lab. 36).

Alunogene: In brown coal at Tuapeka and Manawatu (Hector).

Ambrite (Amberite): See under Resin.

Amethyst: See under Quartz.

Amianthus (*a fine silky form of Asbestos*): Arrow district (Skey, Lab. 31).

Amphibole: A common constituent of igneous rocks. See Hornblende, Actinolite, Nephrite, Riebeckite, Smaragdite, Arfvedsonite, Tremolite, Asbestos, &c.

Analcite (Analcime—*One of the Zeolites*): In volcanic rocks in various districts. Reported from Dunedin (Marshall 1906); Awatere Valley (Thomson 1913).

Andalusite: In schist, Parapara district (Bull. 3); and in mica-schist, Taramakau Valley and Arahura River, Westland (Sollas 1906).
See also Chiastolite.

Andesine: A very general rock constituent in all the intermediate to basic igneous rocks of New Zealand.

Anglesite (*Lead Sulphate*): Old Champion Mine, Te Aroha (Park 1912).

Ankerite: Mount Jumbletop and elsewhere in North Westland (Bull. 1 and 6), and Thames (Bull. 10).

Anomite (*var. of Mica*): Probably in basanite near Mount Cargill, Dunedin (Bartrum, Trans. 41).

Anorthite (*one of the Feldspars*): Is found frequently in the more basic igneous rocks.

Anorthoclase (*one of the Feldspars*): Rock constituent. Abundant in the alkaline rocks of Dunedin (Marshall 1906); described also from Westland (Smith, Trans. 40), and from Campbell Island (Marshall, Trans. 41).

Anthophyllite: Dun Mountain, Nelson (Davis, Rep. G.S. 1871).

Anthracite: See under Coal.

Antigorite (*a variety of Serpentine*): Griffin Range, North Westland (Bonney 1908); slopes of Mount Bowen and Upper Hokitika River, North Westland (Bonney and Bull. 6); Dun Mountain (Davis and Cox); Parapara (Bonney 1908, G.S. 1913).

Antimonial Ochre: See Cervantite.

Antimonite: See Stibnite.

Antimony (*Native*): Great Barrier Island (Skey, Lab. 25). The occurrence of native antimony in New Zealand requires verification. Metallic antimony forwarded from Ross, Westland, in 1911 was determined by the Dominion Laboratory as a smelted sample (Lab. 45).

Antimony-ores: See Stibnite, Cervantite, &c. Antimony-ore has been mined at Hindon, Otago; Endeavour Inlet, Marlborough; Waiheke Island, Auckland, and several other localities.

Apatite: General minor rock constituent—in almost all igneous rocks. For earthy and massive forms see under Phosphates.

Apophyllite: In amygdaloids at Rangitata, Canterbury (Haast 1865); Turnagain Point (Hector, Rep. G.S. 1892).

Aragonite (*Carbonate of Calcium*): Chiefly in cavities in basic volcanic rocks—Coromandel (Bull. 4); Thames (Hutton 1869); Whangaroa (Cox, Trans. 15, and Skey, Lab. 10); Waitaki and Oamaru district (Skey, Lab. 5 and 13, and Park); East Cape, from hot springs (Hector); Collingwood, from caves (Hector); Banks Peninsula (Haast); Dun Mountain district, as gangue of copper-ore (Bull. 12); Dunedin and Caversham (Liversidge, Trans. 10); Quartz Ranges, Collingwood (Cox, Rep. G.S. 1882).

Arfvedsonite (*var. of Amphibole*): In trachyte, Banks Peninsula (Speight, Trans. 40).

Argentite (*Silver Glance*): Port Charles, Coromandel (Bull. 4); Monowai Claim, Thames (Bull. 10); Old Silver Crown Claim, Thames (Davis, Rep. G.S. 1871); many of the mines of Te Aroha, Karangahake, Waihi, and other parts of the Hauraki Goldfield. Is an important ore in several of these mines. Occurs also at Puhipuhi and at Great Barrier Island in some quantity.

Arsenic (*Metallic*): Kapanga Mine, Coromandel (Hector 1864, and Skey, Lab. 14); Thames Goldfield (Skey, Lab. 3, and Hector, Trans. 2).

Arsenolite (*Arsenic Trioxide*): Coromandel (Maclaren 1900).

Arsenopyrite (Mispickel—*Sulpharsenide of Iron*): Milford Sound (Hector 1865); Thames (Pond, Mines Rep. 1887, p. 58); Collingwood (Cox, Trans. 14); Malvern Hills (Hector, Rep. G.S. 1892); Waipori (Hector, Rep. G.S. 1892); Coromandel (Bull. 4); Boulder River, Parapara (Bull. 3); various mines at Waihi and Waitekauri (Bull. 15); Carrick Range, Otago (Maclaurin, Lab. 39); Paringa River (Skey, Lab. 11); Saddle Hill, Dunedin; Lake Wakatipu (Cox, Trans. 14); Langdon's, Greymouth (Cox, Trans. 14); Blackwater Mine, Reefton (McPadden); George Sound (Don); Mount Rangitoto, Westland (Don); reported by Skey (various Lab. reports) from—Paparoa Ranges, Greymouth; Preservation Inlet; Mount Rochfort. near Westport; Wairau Valley, Marlborough.

Artesian Water: See Water—Artesian.

Asbestos (*commercial term for fibrous varieties of Amphibole and Serpentine*):—

Amphibole: Rough Wainihinihi, North Westland (Bull. 1); Mounts Jumbletop, Inframeta, and Bowen, North Westland (Bull. 6).

Serpentine: This fibrous variety is termed *Chrysotile*. Occurs in most localities where massive serpentine is found. Richmond Flat, Parapara (Bull. 3); near Cromwell at the source of the Springburn, Otago (Bull. 5); Dun Mountain district (Bull. 12); Milford Sound, Takaka, Mount Arthur, and Mount Pisa, Otago (Hector, Rep. G.S. 1892); Arrow district (Skey, Lab. 31); Red Hill, Big Bay (Skey, Lab. 22); French Pass (Maclaurin, Lab. 40). Some of the above occurrences of so-called chrysotile may be true asbestos (amphibole), from which mineral the chrysotile differs only in the presence of water.

Asbolite (Earthy Cobalt—*contains Oxide of Cobalt*): Auckland district (Pond, Trans. 10); Rapaka, Bay of Islands (see Park 1910).

Augite: A common constituent of igneous rocks, occurring particularly in the widespread andesites and basaltic rocks.

See also Diopside, Diallage, Ægirine (Ægirite).

Awaruite (*Iron-Nickel alloy*): Jackson's Bay district (Skey, Lab. 21); in sands of Gorge River and other streams from the Red Hill Range, North-west Otago (Marshall 1912). The alloy is apparently an impregnation in serpentine (Marshall 1912).

Azurite (*Blue Hydrous Oxy-carbonate of Copper*): Has been reported from Coromandel (Bull. 4); Mount Radiant (Bull. 11); Dun Mountain (Bull. 12); Parapara district (Bull. 3); Great Barrier Island (Hector, Rep. G.S. 1892). It occurs in many of the localities where chalcopyrite (copper-pyrites) is present.

Barite (*Sulphate of Barium*): Waikouaiti (Mantell 1852); Akiteo (Hector 1867); Sherry Valley, Nelson (Maclaurin, Lab. 44); Wade (Rep. G.S. 1884); near Millerton and Coalbrookdale, Westport (G.S. Ann. Rep. 1912); Thames (Skey and Bull. 10); Skey in Lab. reports records occurrences at Baton River, Nelson; Tenui; Collingwood; East Cape; Makara; Red Island, Cape Kidnappers; Shag Valley, Otago; Te Arai, Kaipara; Pahi, Kaipara; Waitaki Ranges; Whangaroa; Herbertville. Park (1910) mentions Paonui Point, near Napier; Puriri; and Waihi.

Barkevikite (*var. of Amphibole*): A not uncommon constituent of camptonites and other igneous rocks.

Basanite: See Lydian Stone, under Quartz.

Bastite (Schiller Spar—*altered Enstatite*): West Coast, South Island (Hector 1865); Dun Mountain district (Bull. 12); North Cape (Marshall 1909).

Bathvillite (*allied to Torbanite*): Mangonui (Skey, Lab. 31).

Berthierine (*Hydrous Silicate of Iron-protoxide*): Dun Mountain district (Skey, Lab. 9); D'Urville Island (Skey, Lab. 10); Tawa Flat, Wellington (Skey, Lab. 15).

Beryl: Dusky Sound (Skey, Lab. 16); Paterson's Inlet, Stewart Island (Skey, Lab. 24, and McKay, Rep. G.S. 1890).
See also Emerald.

Biotite (*Magnesium-Iron Mica*): A very common constituent of granitic and gneissic rocks. Large crystal plates at Port Pegasus, Stewart Island.

Bismuth (*Native*): Coromandel—in gravels of Wade Creek (Bull. 4); Owen River, Nelson (Hector and Skey, Lab. 22 and 23); Westport (Fry, Mines Record, vol. x, p. 11).

Bitumen: Cast up on south and east coasts of New Zealand (Skey, Lab. 3) and near Mokau (Bull. 14); Poverty Bay (Liversidge, Trans. 10).

Blacksand: See Ironsand; Iserine; Magnetite; Ilmenite.

Bloodstone (*var. of Quartz*): See under Quartz.

Bog-iron Ore: See Limonite.

Bole (*var. of Halloysite*): Lyttelton Tunnel (Haast; see Cox, Trans. 15).

Bornite (*Sulphide of Copper and Iron*): Wilberforce Reefs, North Westland (Bull. 1); Coromandel (Bull. 4); Whangaroa (Bull. 8); Mount Radiant (Bull. 11); Thames (Bull. 10); Cape Runaway; Kawarau River, Otago (Hector, Rep. G.S. 1892); Kawau (Hector, Trans. 2); Dunstan, Otago (Liversidge); Parapara (Bull. 3).

Boron: In mineral waters from Rachel Spring, Rotorua, and four miles and a half south of Te Kuiti (Maclaurin, Lab. 46).
See also Tourmaline.

Bournonite (*Sulph-antimonite of Lead and Copper*): Rolling River, Wangapeka (Hector, Rep. G.S. 1892); Central Otago (Finlayson, Trans. 41).

Bowenite (*var. of Serpentine*): By the Maoris called Tangiwai, and used for ornamental purposes. In veins in talc at Milford Sound (Cox, Trans. 15); Jackson's Bay (Cox, Trans. 15); Pounamu formation in North Westland (Bull. 1); Mount Jumbletop, Westland (Bull. 6).

Braunite (*chiefly Protoxide of Manganese*): Mangapai, Auckland (Skey, Lab. 26); Malvern Hills (see Cox, Trans. 14); Wellington and Bay of Islands (Skey—Park 1910).

Bromyrite (*Silver Bromide*): Probably in Talisman Mine, Karangahake (McDuff).

Bronzite (*Ferriferous Enstatite*): West Coast (Hector 1865); recorded also by Davis (Rep. G.S. 1871) from the Dun Mountain, Nelson, but this identification is probably incorrect (see Bull. 12).

Brookite (*Titanium Dioxide*): In dolerite at Otepopo (Hector 1865).

Bytownite (*var. of Feldspar*): Constituent of basic and ultra-basic igneous rocks.

Calamine (*Carbonate of Zinc*): Tararu Creek, Thames (Skey, Lab. 6 and Rep. G.S. 1871).

Calc-spar: See Calcite.

Calcite (*Carbonate of Lime*): This mineral, the essential constituent of limestone, marble, &c., is abundant throughout New Zealand. Hector records as localities: Tokatea Range (? Coromandel); in Tertiary rocks of Otago as Dogtooth Spar; Nelson, in limestone caves; Canterbury as Iceland Spar (Haast 1864); Dunedin; Seacliff, near Waikouaiti; Cape Rodney; Tararu Creek, Thames, as smoke-coloured calcite. Maclaren (1900) mentions the following varieties from Coromandel: Iceland Spar, Dogtooth Spar, Nailhead Spar, Argentine. See Liversidge's list (Trans. 10), G.S. Bulletins, &c.

See also Marble; Chalk.

Cancrinite: Frequently associated with nepheline in Dunedin alkaline igneous rocks; also in similar rocks from gravels of New River, Westland (Smith, Trans. 41).

Carnelian (*var. of Quartz*): See under Quartz.

Cassiterite (*Dioxide of Tin*): All the occurrences, with the exception that at Stewart Island the mineral is found also in veins, are of stream-tin. It has been found at Lankey's Gully, Reefton; Hope Valley, Buller River; Wairenga, Kaihika, Poverty Bay (doubtful); Campbell Island: these occurrences are recorded in Lab. reports. Also reported from Montgomery's Claim, Blackball (Bull. 13); Dusky Sound; foot of Mount William Range, Westport (Handbook of N.Z. Mines, 1887); Mackley River, Westport (*R*); Tuapeka (Park 1910); in auriferous gravels to the north-east of Westport (S. Fry); Ten-mile Creek north of Greymouth (*R*). Mining operations are in progress at Stewart Island. See Description of Exhibits, p. 24.

Cerargyrite (Horn Silver—*Silver Chloride*): Maratoto and Waihi (Skey, Lab. 25 and 21); Waikoromiko Valley, Coromandel (Bull. 4); Talisman Mine, Karangahake (McDuff); Thames, Waitekauri, Te Aroha, Puhipuhi and Great Barrier Island (Park 1910). The mineral in all cases is in the gold-silver-veins of the districts.

Ceria, &c.: See Monazite, Rare Earths.

Cerussite (*Lead Carbonate*): In the upper portions of galena-bearing veins at Te Aroha, especially at the Tui Mine, where masses of many tons weight were found (Park 1910).

Cervantite (*Oxide of Antimony*): Widely distributed encrusting stibnite. Localities reported are Thames (Bull. 10); Langdon's Creek, Greymouth (Bull. 13); Otaio, Canterbury (Skey, Lab. 30); Hindon in Otago, and elsewhere.

Chabazite (*one of the Zeolites*): In small cavities in volcanic rocks. Dunedin (Hector 1865); Banks Peninsula (Haast 1865); Helenburn, Otago (Cox, Trans. 15).

Chalcanthite (Blue Vitriol—*Hydrous Cupric Sulphate*): Stalactites in old drives at Thames (Bull. 10) and Waihi (Bull. 15).

Chalcedony (*var. of Quartz*): See under Quartz.

Chalcocite (Copper-glance—*Cuprous Sulphide*): Whangaroa (?), (Bull. 8); Mount Radiant (Bull. 11); Dun Mountain district (Cox. Trans. 14); Jackson's Bay (Skey, Lab. 11); D'Urville Island (Park 1910).

Chalcopyrite (Copper-pyrites): Numerous localities. Cox (Trans. 14) mentions the following: Kawau Island; Great Barrier Island; Lake Wakatipu; Waipori; Paringa River, Westland; Collingwood; Waiomo, Thames; Dusky Sound. Other localities are—Dun Mountain, Nelson (Bull. 12); mines of Waihi (Bull. 15) and of Thames (Bull. 10); Mount Radiant (Bull. 11); Karangahake; Te Aroha; Coromandel (Bull. 4); Moke Creek, Wakatipu (Hector, Rep. G.S. 1892); Maharahara; Pahi (North Auckland); Whangaroa (Bull. 8); several localities in Westland in the Pounamu formation (Bull. 1 and 6); Fourteen-mile Creek, north of Greymouth; Lake Ohau and Moorhouse Range, Canterbury (Park 1910); Opotiki; Cape Runaway.

See also list of localities under Copper-ores.

Chalcotrichite (*var. of Cuprite*—Plush Copper Ore—*Cuprous Oxide*): Champion Copper-mine, Aniseed Valley, Nelson (Cox, Trans. 16, p. 448).

Chalk: Oxford, Canterbury (McKay). The Amuri limestone, which has a wide distribution throughout New Zealand, has in many places a chalky facies.

Chiastolite (*var. of Andalusite*): Slate River, Collingwood (Cox, Trans. 15); head of Anatori River, in schist (Isaacson—see Bull. 3).

Chlorite: A common constituent of altered rocks. See also Clinochlore, Delessite, Chlorophæite.

Chloropal (*Hydrated Silicate of Ferric Iron*): Otago (Liversidge).

Chlorophæite (*var. of Chlorite*): In cavities in igneous rocks. Mount Somers, Rangitata, and Malvern Hills (Haast).

Chromglimmer (? Fuchsite): Dusky Bay (Skey, Lab. 16).

Chrome Mica (? Fuchsite): North Westland (Bull. 1); Cox (Trans. 15) records Lake Wakatipu and Dusky Sound. In the North Westland occurrence it is associated with ruby in the material called Goodletite.

See under Ruby.

Chrome Ochre: Nelson, in combination with chromite (Hackett 1861).

Chromite (*Chromium-Iron-oxide*): D'Urville Island (Hochstetter); Ben Nevis, near Nelson (Park 1910); Red Mountain, Otago (Hector 1865); Dun Mountain and Aniseed Valley (Hochstetter, and Bull. 12); Pounamu formation, North Westland (Bull. 1); Mikonui district, North Westland (Bull. 6); in gravels, Greymouth (Bull. 13), and near Westport. Other localities recorded in Lab. reports are: Croixelles, Nelson; Lake Harris, Otago; Martin's Bay; Pelorus Valley; Takaka; Lake Wakatipu; and Whangamoa, Nelson. It occurs as small grains in ultrabasic rocks of south-west Otago, Nelson, and North Westland (see for example Marshall, Trans. 37).

Chrysoberyl (*Beryllium Aluminate*): Stewart Island (Skey 1889; see Hector, Rep. G.S. 1892).

Chrysocolla (*Hydrous Copper Silicate*): "Mineral Belt," Nelson (Cox, Trans. 14 and Bull. 12); Mount Radiant (Bull. 16); Karangahake (G.S. 1912); Waihi (Morgan); Lyell, Buller (Skey, Lab. 21); Paterson's Inlet (Skey, Lab. 17); Mount Rangitoto, Westland.

Chrysolite: See Olivine.

Chrysoprase (*var. of Quartz*): See under Quartz.

Chrysotile (*Fibrous Serpentine*): North Westland (Bull. 1); Parapara district (Bull. 3); Dun Mountain (Bull. 12). See also Asbestos.

Cinnabar (*Sulphide of Mercury*): Waiohine Ranges, Wellington (McKay, Rep. G.S. 1888); Cox (Trans. 14) gives the following occurrences: Obelisk Ranges; Potter's Gully, Dunstan; Serpentine Valley and Waipori, Otago—all in alluvial deposits; also at Ohaeawai hot springs, North Auckland. Other localities are: North Auckland at Kaeo, Kirikiri River, Puhipuhi, and Waimate Bay; Te Aroha; Taipo, North Westland—all these are from Lab. reports by Skey and Maclaurin; Mackaytown, Karangahake (Bull. 15); Dun Mountain (*R*) (Bull. 12); Coromandel, Thames, &c. (Pond, Mines Rep., C.-3, 1887); several places in the Waihi district (Bull. 15); near Waipori, Otago, in vein; Waiohine River (Skey, Lab. 18); Nevis and Nokomai district, Otago (Finlayson, Trans. 41).

See also Description of Exhibits, p. 24.

Clay: Clays for most purposes are abundant in New Zealand, though material suitable for the higher classes of pottery is not known to exist in quantity.

See also Kaolin; Fuller's Earth.

Clinochlore (*var. Chlorite*): This mineral is probably widely distributed in various rocks, but has not been often actually described as such. Schists of Otago (Marshall 1912); Parapara (Thomson, Trans. 42).

Clinozoisite (*var. of Epidote*): (?) Granites and gneisses in the Mikonui district, Westland (Bull. 6); similar rocks in the Mount Radiant district (G.S. 1913); in amphibolite at Waikawa (Thomson, Trans. 42).

Coal: Every variety of coal, from anthracite to lignite, occurs in New Zealand. It is not possible to enumerate every locality in which coal of some kind is found, but the chief occurrences are,—

Anthracite: Fox River; Malvern Hills; Acheron River; Cabbage Bay, near Coromandel (Bull. 4, p. 57).

Semi-anthracite: Paparoa Coal-mine, near Greymouth.

Bituminous Coal: Collingwood; Westport; Greymouth. Occurs also in various parts of central Nelson, in Urewera country (extent unknown), at Cabbage Bay (Bull. 4, p. 57), Malvern Hills (brown coal altered by intrusions of igneous rocks); Paringa River, South Westland (Lab. 15, pp. 29–30).

Glance or *Pitch Coal:* Several parts of North Auckland, including Kawakawa, Hikurangi, Ngunguru, and Kiripaka districts; Reefton; Greymouth (Point Elizabeth State Coal-mine).

Brown Coal and *Lignite:* The almost innumerable localities include: Waikato district; Raglan; Mokau River; Retaruke River; West Wanganui (Westhaven); Heaphy River; Inangahua Valley; Shag Point; Kaitangata; various localities in Wellington and

Hawke's Bay; Charleston and Brighton (south of Westport); Malvern Hills; Mount Somers; Waihao; Albury; Ngapara; Papakaio; Green Island; Milton; Stirling; Roxburgh; Alexandra; Clyde; Cromwell; Bannockburn; Gore; Pukerau; Waikaia; Forest Hill; Nightcaps; Orepuki; Preservation Inlet (Coal Island); Tangarakau; Ohura Valley.

Full information concerning coal-mining in New Zealand may be obtained from the annual reports of the Mines Department, entitled "Papers and Reports relating to Minerals and Mining." General and geological information is given in various geological reports and bulletins.

See also Description of Exhibits, pp. 15–17.

Cobalt: Traces in copper-ore at Karamea (Bull. 11) and Dun Mountain (Bull. 12); in manganese-ore at Bay of Islands (Skey, Lab. 13), Kawau Island (Skey, Lab. 14), New Plymouth (Maclaurin, Lab. 38), Taieri Mouth (Morgan and others), Waihi and Waitekauri (Morgan); and in pyrrhotite at Richmond Hill, Parapara (Skey, Lab. 13 and 22). In many of these occurrences the cobalt is probably in the form of an oxide.

See also Asbolite and Erythrite.

Cobalt Bloom: See Erythrite.

Copalite: See Resin—Mineral.

Copper (*Native*): Cox (Trans. 14) records—Moke Creek, Wakatipu; Serpentine Belt, Aniseed Valley and Dun Mountain, Nelson; Great Barrier Island; Perseverance Mine, Collingwood; Manukau North Head. Hector (Rep. G.S. 1892) adds Thames (see also Bull. 10) and Perseverance Mountain (Skey). Other localities are—Golden Ridge, Collingwood (Bull. 3); Coromandel (Bull. 4); Whangaroa (Bull. 8); Karori (Skey, Lab. 31); Pahi, Kaipara (Skey, Lab. 20); Whangamoa district, Nelson (Skey, Lab. 33); Kawau Island (Baker, Trans. 33); Maharahara (Hector); in syenitic rock, Dusky Sound (Don).

Copper Glance: See Chalcocite.

Copper-ores: The following localities are recorded by Lab. reports: Batley, Kaipara; Bay of Islands; Bull's; Dannevirke; D'Urville Island; Dusky Sound; Great Barrier Island; Lake Hayes, Otago; Hokianga; Jackson's Bay; Kaeo district, North Auckland; Maharahara; Malvern Hills, Canterbury; Manawatu Gorge; Mangatapu Survey District, Nelson; Mangonui; Parapara; Petote Creek, Coromandel; Pirongia Range; Mount Radiant; Ruahine Ranges; Te Anau; Waipori; Waitahuna, Otago; Whangamoa, Nelson; Whangaroa; Woodville (Maharahara).

See also Chalcopyrite, &c., and Description of Exhibits, pp. 22–23.

Copper-pyrites: See Chalcopyrite.

Copperas: See Melanterite.

Coprolites: Wellington Heads, in a cave (Maclaurin, Lab. 42); Kaikoura district (Skey, Lab. 32); Swampy Hill and Kaikorai Valley, near Dunedin; near Oamaru (Aston).

See also Phosphates.

Cordierite: See Iolite.

Corundum (*Sesquioxide o, Aluminium*): Collingwood (Skey, Lab. 32); Parapara (Skey, Lab. 27); Upper Moutere, Nelson (Skey, Lab. 30); Rimu and Kanieri, Westland (Bull. 1 and 6).
See also Ruby, Sapphire.

Cossyrite (*var. of Amphibole*): See Ænigmatite.

Cotunnite (*Lead Chloride*): Komata, Thames Goldfield (Park 1910).

Covellite (*Cupric Sulphide*): D'Urville Island (Cox and Hector, see Rep. G.S. 1892); Whangaroa (Bull. 8); Karangahake (G.S. 1912); Mount Radiant (Bull. 11); Red Hill, Big Bay (Skey, Lab. 22); Te Aroha (Skey, Lab. 17); Wangapeka (Skey, Lab. 17).

Cuprite (*Red Oxide of Copper*): Dun Mountain district (Cox, Trans. 14, and Bull. 12); Mount Radiant (Bull. 11); Thames (Hutton 1867); Bligh Sound and Lake Te Anau (Hector); Waihi Mine (Morgan); D'Urville Island (Park 1910).

Delessite (*var. of Chlorite*): Andesite at Thames (Sollas 1906); Mount Somers and Malvern Hills, Canterbury (Haast).

Dermatin: Dun Mountain and West Coast Sounds (Davis—see Hector 1892).

Diallage (*var. of Pyroxene*): In ultrabasic plutonic igneous rocks, Darran Mountains (Marshall 1912); Whangaroa (Bull. 8); Dun Mountain (Bull. 12); Mount Torlesse and Upper Rakaia, Canterbury (Haast); Martin's Bay; Kakapo Lake and Mount Arthur (Hector, Rep. G.S. 1892); Milford Sound (Marshall, Trans. 39); Lake McKerrow and diorites on the West Coast (Hector—Park 1910); Kakanui River (Thomson, Trans. 38).

Dialogite (Diallogite): See Rhodochrosite.

Diatomaceous Earth: Mangonui (Skey, Lab. 30); Hokianga (Maclaurin, Lab. 40); Pakaraka, Bay of Islands (McKay, Trans. 23); Makaretu, Hawke's Bay (Trans. 22, p. 551); Weston, Oamaru (De Latour, Trans. 21); Lake Sumner, North Canterbury (Inglis, Trans. 15); Middlemarch (workable); Akaroa (workable); Green Island, Dunedin; Cust Valley, North Canterbury; Amberley; New Brighton; Macintosh Bay, Banks Peninsula; Whangarei; Cabbage Tree Swamp, Auckland; Foxton (Skey, Lab. 9).

Diopside (*var. of Pyroxene*): In dunite at the Dun Mountain, Nelson (Bull. 12) and in peridotite at Wakatipu district (Marshall); at Kakanui River, North Otago (Thomson, Trans. 38); in dunite at Milford Sound (Marshall, Trans. 37); in inclusions in basalt, Leith Valley, Dunedin (Marshall 1906).

Dioptase (*Silicate of Copper*): Thames (Skey, Rep. G.S. 1871); Dun Mountain (Bull. 12, and Cox, Trans. 14); Whangamoa, Nelson (Skey, Lab. 33).

Disthene: See Kyanite

Dolomite (*Carbonate of Calcium and Magnesium*): Associated with ultrabasic rocks, North Westland (Bull. 1 and 6); Collingwood (Skey, Lab. 8); Malvern Hills (Park 1910). Dolomitic rocks occur in the Collingwood (Bull. 3), Rolling River (Nelson) (Skey, Lab. 20), and Gisborne (Bull. 9) districts, but no massive deposits of sufficient purity to be utilized for basic lining of steel .urnaces are yet known. Dolomitic rock (concretionary) is also reported from Taranaki (Lab. 11, p. 15). The variety pearlspar occurs at Thames.

Domeykite (*Copper Arsenide*): Whakapirau (Skey, Lab. 31).

Dopplerite: Waiapu, formed as a surface deposit by the oxidation of exuded petroleum (Hector 1874). A similar substance has been observed in southern Hawke's Bay, a few miles from Weber. Skey reports "mineral grease" from Waipara [? error in name], North Island (Lab. 16, pp. 56–57).

Dufrenoysite (*Sulpharsenite of Lead*): Great Barrier Island (Skey, Lab. 3), occurring with galena in a diorite (Hutton).

Elaterite: Pieces on the East Coast (Marshall 1912); Kawau Island (Hector 1865); Poverty Bay (Liversidge, Trans. 10).

Electrum (*Natural alloy of Gold and Silver*): Occurs in all the auriferous lodes of the Hauraki Peninsula.
See also Gold.

Emerald (*var. of Beryl*): Dusky Sound (Skey—Cox, Trans. 15).

Enargite (*Sulpharsenate of Copper*): Karaka Creek, Thames (Bull. 10).

Enstatite: Rock constituent. Lake Johnston, Queenstown district (Bull. 7); in olivine rocks near Milford Sound and Nelson (Marshall 1912); North Cape (Marshall, Trans. 41). The mineral has also been recorded from andesites of the Hauraki Peninsula (Hutton, Aus. Ass. Adv. Sc. 1889 and others); Sollas (1906) does not positively determine it in connection with the large number of rocks he described from there and elsewhere in New Zealand, and it is possible that the identifications are incorrect.
See also Hypersthene.

Epidote: Rock constituent. Common in most parts of New Zealand. Granites on west coast of Otago (Hector); Wairarapa, in massive form (Hector); diorites, &c., Mount Torlesse, Canterbury (Haast) and Westport district (G.S. 1913); North Westland, in magnesian rock belt (Bull. 1 and 6), &c.

Epsomite (*Hydrated Sulphate of Magnesium*): Abundant in disused mine-workings at Coromandel (Bull. 4), Thames (Bull. 10) and Waihi (Bull. 15); Otago, as efflorescence (Hector 1865); efflorescence on clays in volcanic regions (Park 1910).

Erubescite: See Bornite.

Erythrite (Erythrine or Cobalt Bloom): Schists and gneisses on the West Coast (Hector 1865).

Fayalite (*var. of Olivine*): Mahunga, Auckland (Skey, Lab. 10); Skippers, Otago (Skey, Lab. 15); Whangaroa Harbour (Skey, Lab. 10); in schist in Otago (Skey—see Park 1910). It is probable that some of these determinations should be referred to Olivine.

Feldspar: See Albite, Oligoclase, Andesine, Labradorite, Bytownite, Anorthite, Orthoclase, Anorthoclase, Perthite, Microperthite, Valencianite, Sanidine, &c.

Flint: Kaipara, as bands in limestone (Park 1910); also at Amuri Bluff, Clarence Valley, and elsewhere in Canterbury, in the Amuri limestone.

Fluorite (Fluor Spar—*Fluoride of Calcium*): Baton River, Nelson (Skey, Lab. 26); Stewart Island (Skey, Lab. 24); Wangapeka, Nelson (Skey, Lab. 24), West Otago (McKay 1889); Herbertville (?) (Skey, Lab. 23).

Fuchsite (*Chrome Mica*): Parapara (Bull. 3); Rimu district, Hokitika, probably in ruby rock (so-called Goodletite), (Bull. 6); in schist at Moonlight Creek, Lake County, Otago (Sollas 1906).
See also Chrome Mica.

Fuller's Earth : In rhyolitic rocks at Waikino (Bull. 15); Gore, Otago (Maclaurin, Lab. 44); Great Barrier Island (Cox, Trans. 15). These occurrences and others that have been reported need practical tests as to their suitability for use as fuller's earth.

Gahnite (Zinc Spinel): Stewart Island (Skey, Lab. 24).

Galena (*Sulphide of Lead*): Richmond Hill, Parapara (Cox, Trans. 14, and Bull. 3); Coromandel (Bull. 4); Mount Rangitoto, Westland (Cox, Trans. 14 and Bull. 6); Whangaroa (Bull. 8); Te Aroha; Mount Radiant (Bull. 11); Thames (Cox, Trans. 14 and Bull. 10); mines at Waihi and Waitekauri (Bull. 15); Kaimanawa Ranges (Crawford), Great Barrier Island (Cox, Trans. 14); Barewood and Bendigo, Otago.
See also Lead-ore.

Garnet : West Otago, in gneiss (Hector); South-west Otago (Marshall, Trans. 39); Dunedin, in basalts (Hector); schists of Parapara (Bull. 3); common in schists and gneisses of Westland (Bull. 1 and 6) and occasionally in granite; Westport district (G.S. 1913). The mineral is frequent in beach sands of Westland, Stewart Island, and elsewhere; when red in colour and of a transparent nature is commonly called "ruby," but is quite distinct from the true or Oriental ruby.
See also Grossularite, Pyrope, Almandine, Uvarovite.

Gas—Natural (*Methane, &c.*): Hanmer; Taranaki, near New Plymouth and elsewhere; Nuhaka; Gisborne to East Cape district; Ure River, Marlborough; Aohanga; Blairlogie, near Masterton; Waipatiki, near Weber, and several other places in eastern Wellington. Carbon dioxide is found in bores at Kotuku, Greymouth, and with methane in artesian wells at Christchurch and elsewhere.

Genthite (*Hydrous Nickel-Magnesium Silicate*): Near Tapu, Thames (Park, Trans. 26); in serpentine from Wade, Hoteo, and Mahurangi, Auckland (Pond and Skey—see Park 1910).

Geocronite (*Sulpharsenite of Lead*): Collingwood (Skey, Lab. 23).

Gersdorffite (Nickel Glance—*Sulpharsenide of Nickel*): Richmond Hill, Nelson (Skey, Lab. 22).

Geyserite : Deposit from hot springs; abundant in the Taupo and Rotorua districts.

Glauber Salts : See Mirabilite.

Glauconite : Abundant in the "greensands" below Tertiary and Cretaceous (?) limestones—*e.g.*, Gisborne district, Hawke's Bay, eastern Wellington, North Westland, North and South Canterbury, North Auckland, &c.

Gmelinite (*one of the Zeolites*): Dunedin (Hector 1865).

Göthite (*Hydrated Sesquioxide of Iron*): In cavities in the Parapara limonite (Bull. 3).

Gold :—

Alluvial : Alluvial gold occurs in most parts of New Zealand. Most of the richer alluvial deposits have been worked in Otago, Nelson, and Westland.

Lode Gold : Quartz and other lodes carrying gold occur in the Tertiary volcanic rocks of the Hauraki Peninsula and in almost all parts of New Zealand where pre-Cretaceous sedimentary rock are exposed.

For description of the gold-mining industry see Description of Exhibits, pp. 6-13.

Graphite: In graphitic schists and slates of Parapara and Collingwood districts (Bull. 3 and elsewhere); Cromwell district, in veins in schist (Bull. 5); Mount Potts, Canterbury; Reefton; near Thames (Morgan, in MS.); near Wellington (Crawford, Trans. 1), on the Petone Road, and in the Rimutaka Mountains, in the valleys of the Waiohine, Ruamahunga, Waikanae, Waingawa, Akatarawa, and Otaki Rivers. Recorded also in various annual reports of the Laboratory at Avondale, Marlborough; Clyde, Otago; Dusky Sound; Mount Egmont; Nelson; Orari Gorge, South Canterbury; Pakawau, Nelson; Rangiahua, Auckland; Waimate, South Canterbury; Paparoa Range or neighbourhood. Thin flat veins at Pakawau (near Collingwood) in schist were worked prior to 1866 (Hector).

Greenstone: See Pounamu, Tangiwai, Nephrite, Bowenite, &c.

Grossularite (*Lime-Alumina Garnet*): Forms a large proportion of the rock named Rodingite by Bell, Clarke, and Marshall (Bull. 12) which occurs at the Dun Mountain, Nelson, and was first noted as a grossularite-gabbro by Marshall (Trans. 40, pp. 320-322).

Gypsum (Selenite—*Hydrous Calcium Sulphate*): Mount Ida, Otago; Awamoko (Liversidge, Trans. 10); Lake Wakatipu (Liversidge, Trans. 10); Komata, Hauraki Peninsula; White Island (Liversidge, Trans. 10); Moeraki (Liversidge); Parapara iron-ore in cavities (Bull. 3); mines at Thames (Skey, Lab. 20 and Rep. G.S. 1871; also Bull. 10); Kaikoura district and Lower Waipara Gorge (G.S. 1913); near Waipori (Finlayson, Trans. 41). Additional localities (Cox, Trans. 15) are—Malvern Hills; Kaitaki Hills (Patua Range); and Waihao.

Hæmatite (*Sesquioxide of Iron*): Cox (Trans. 14) gives the following occurrences: Near Lake Wakatipu, in persistent band; Maori Point, Shotover; Dunstan, Otago; Otamataura Creek, Collingwood; Whangaroa. Additional localities are—Roaring Meg and Mount Pisa, Otago (Park 1910); Greymouth (Skey, Lab. 18); Pahua, Wairarapa (Skey, Lab. 18); Wharekawa, Auckland (Skey, Lab. 10). A micaceous variety is recorded by Marshall (1912) from near Queenstown.

Halite (Rock Salt—*Chloride of Sodium*): Salt springs occur in Masterton, Tane, and Gisborne districts. There is also salt water in petroleum bores at Kotuku, near Greymouth, and at New Plymouth.

Halloysite: Dunedin (Hector 1865, Liversidge, Trans. 10); Whangaroa (Cox, Trans. 15); Karangahake (Skey, Lab. 19); Riwaka, Nelson (Maclaurin, Lab. 42); Scinde Island, Napier (Skey, Lab. 7); Bay of Islands and Drury and Hunua Range (Skey—see Park 1910).
See also Smectite.

Hauerite (*Manganese Disulphide*): Lake Wakatipu (Skey, Lab. 15 and 16); Collingwood (Cox, Trans. 14).

Hausmannite (*an Oxide of Manganese*): Rolled pieces in the Selwyn River gravels (Haast 1865).

Hectorite: Described by Cox as a new mineral of the serpentine group from the Dun Mountain district, Nelson (Trans. 15, p. 409).

Hedyphane: Coromandel (Maclaren 1900).

Hercynite (Iron Spinel): Highlay Hills (Skey, Lab. 25).

Hessite (*Silver Telluride*): Waitangi and May Queen Mines, Thames (Bull. 10); Maratoto mines (Bull. 15); Te Aroha (probable) (Skey, Lab. 22).

Heulandite (*one of the Zeolites*): In felsite porphyry, Canterbury (Haast 1865); andesite, Matawai Stream, Coromandel (Sollas 1906).

Hornblende: Rock-constituent. Very common in almost all the igneous and in some of the metamorphic rocks of New Zealand.

Hyalite (*var. of Opal*): Dunedin, on the surface of some volcanic rocks; Snowy Peak and Malvern Hills, Canterbury (Haast); Clarendon (Andrew).

Hydromagnesite (*Basic Magnesium Carbonate*): Dun Mountain district (Bull. 12); near Karangahake (McDuff).

Hydrophite: See Dermatin.

Hypersthene: Rock-constituent. Common in the andesites of the Hauraki Peninsula and in similar rocks throughout the North Island; in basic and ultrabasic igneous rocks at Malvern Hills (Haast 1865), Dun Mountain (Davis, Rep. G.S. 1871), the West Coast (Hector 1865), and South-west Otago (Marshall, Trans. 39).

Iddingsite (*Iron-rich Serpentine*): In spessartites, &c., at Mount Tapuae-nuku, Marlborough (Thomson, Trans. 45); in camptonitic rocks in the Paparoa Mountains, near Westport, and in gravels of Central Otago (G.S. 1913).

Idocrase: (Vesuvianite—*Basic Calcium-Aluminium Silicate*): Dusky Sound (Cox, Trans. 15).

Idrialite or so-called "inflammable cinnabar" (*an Oxygenated Hydrocarbon associated with Cinnabar*): Near Waimate North (Skey, Lab. 10); Dunstan and Waipori (Park 1910); Ohaeawai hot springs, as rounded grains in alluvium (Hector, Rep. G.S. 1892). The Waimate North locality is probably Ohaeawai.

See also Dopplerite.

Ilmenite (Menaccanite—Titanic Iron-ore—*an Oxide of Iron and Titanium*): Richmond Hill, Parapara (Skey, Lab. 7); Brancepeth, Wairarapa (Skey, Lab. 8); is a common constituent of igneous rocks and of sands derived therefrom. Large deposits of titaniferous sands occur along the Taranaki coast near New Plymouth, Patea, &c., and to a less extent along the west coast of the South Island.

Infusorial Earth (*var. of Tripolite*): See Tripolite; Diatomaceous Earth.

Iodine: Occurs in many mineral waters, especially from cold springs in Wellington, Hawke's Bay, and Gisborne districts.

Iolite (Cordierite): Mokihinui River (Bull. 11); Greymouth district, in gravels (J. P. Smith and G.S. 1912); in contact rocks of the Aorere district, Collingwood (Marshall, Trans. 41).

Iridium: Occurs with platinum at Thames (Pond, Trans. 15) in quartz reef at the Queen of Beauty shaft; near Takaka (Bull. 3).

See also Osmiridium, Platiniridium, &c.

Iron (*Native*): In a few meteorites, *e.g.* one at Wairarapa; in the nickel-iron alloy called Awaruite at Gorge River, north-west Otago.

Iron-ore: See Limonite, Hæmatite, Magnetite, &c., and Description of Exhibits, pp. 19–20.

Ironsand: The so-called ironsands of the New Zealand coast consist mainly of titaniferous magnetite. On the west coast of the North Island, near New Plymouth, Patea, &c., they are so abundant as to constitute an important future source of iron.

See also Magnetite; Iserine; Ilmenite.

Iserine (*var. of Ilmenite*) : Ironsands of the west coast of the South Island (Hector, Park 1910).

Jade : See Nephrite.

Jasper (*var. of Quartz*) : See under Quartz.

Jet (*var. of Coal*) : Maharahara (McKay) ; Westport (S. Fry) ; Methven (Lab. 35).

Kaolin : Occurs with the veinstone of the Waihi and other mines of the Hauraki Peninsula ; Manuherikia ; Arrow River ; Mount Somers ; Collingwood ; Stewart Island ; New Plymouth ; Kakahu, South Canterbury ; Whangarei Harbour ; and in many other localities in various parts of New Zealand recorded by the Lab. reports. Some of these deposits may be of commercial importance.

Kauri-gum (*Kauri-resin*) : Occurs in the North Island north of a line drawn from Kawhia to Tauranga. The digging of kauri-gum is an important industry in New Zealand. See pages 17–19 of Description of Exhibits.

Kerargyrite : See Cerargyrite.

Kermesite (Antimony-blende—Kermes—*Oxysulphide of Antimony*) : Occurs generally with stibnite according to Cox (Trans. 14). Recorded from Tararu Creek, Thames (Skey, Lab. 6) and from Coromandel (Maclaren 1900).

Kieselguhr : See Diatomaceous Earth.

Kyanite (Disthene) : Westland, associated with quartz (Hector, Rep. G.S. 1892).

Labradorite (*var. of Feldspar*) : Very general constituent of intermediate and basic igneous rocks throughout New Zealand.

Laumontite (*one of the Zeolites*) : Nests and veins of this mineral probably occur at Waihi (Morgan, Trans. Aust. Inst. Min. Eng. 1902) in altered andesite.

Lead (*Native*) : Thames (Skey, Rep. G.S. 1871) ; Collingwood (Skey, Lab. 24). These occurrences probably represent lead shot and not native lead. Metallic lead, certainly derived from shot, is common in the Molyneux River, and is saved with gold by the dredges.

Lead Chloride : Waiomo, Thames (Allen, Mines Record, vol. ii, p. 23).

Lead-ores : Have been reported in various Lab. reports from—Buller district ; Collingwood ; Eltham, Taranaki ; Mount Rangitoto, Westland ; Parapara ; Ruahine Ranges ; Stewart Island ; Thames (Monowai, Colorado, Sylvia, and other mines) ; Coromandel ; Waitaha River, Westland ; Wangapeka, Nelson. Lead-ores are also found at Puhipuhi, Te Aroha, in grains in schist and gneiss in Otago and Westland (Park 1910), and in several additional localities mentioned under Galena. No lead mine has yet been developed in New Zealand.

Lepidolite (*Lithia Mica*) : Thomson Sound (Hector), in marble.

Lepidomelane (*var. of Mica near Biotite*) : Schists and gneisses of the West coast (Hector 1865) ; Milford Sound (Cox, Trans. 15).

Leucite : Rock-constituent. In leucitophyre and allied rocks at Dunedin (Ulrich, Marshall, and Bartrum) ; in basalt near Castle Point (McKay).

Leucopyrite (*Iron Arsenide*—Fe_3As_4): Thames (Cox); Collingwood (Cox); Reefton (Cox—see Hector, Rep. G.S. 1892); Langdon's Creek, Greymouth (? Hector 1879—see Bull. 13).

Leucoxene (*an alteration product of Titanic Iron-ore*): Common in many igneous rocks. In some cases siderite has probably been mistaken for this mineral.

Limonite: The localities in which this mineral occurs are too numerous to be included in this list. Deposits of great economic importance occur near Parapara, Nelson, and have been fully described in Bulletin No. 3. See also Description of Exhibits, pp. 19–20. Minor deposits occur near the head of the Waitangi River, at Kerr Point, and elsewhere in North Auckland. There is said to be a large deposit of "brown hæmatite" at Mount Peel, Nelson, but this has not yet been investigated. Several limonitic deposits have been discovered in Otago, but none is known to be of importance.

Lithium: In mineral waters at Te Aroha and elsewhere; also in Lepidolite, which see.

Lithographic Limestone: Campbell Island (Chapman, Trans. 23); Oruru Valley, Hokianga (McKay, Rep. G.S. 1894); Abbey Rocks, South Westland (Cox, Rep. G.S. 1877); Chatham Islands (Hector, Handbook of N.Z., 1879).

Löllingite (*Diarsenide of Iron*—Fe As_2): Victory Mine, near Lyell (determined by Lab.).

Lydian Stone (*var. of Quartz*): See under Quartz.

✓ **Magnesite** (*Carbonate of Magnesium*): Rotorua (Cox 1878); Chatham Islands, massive (Smith—see Hector, Rep. G.S. 1892); Collingwood (Skey, Lab. 13, and Cox, Trans. 15); Milford Sound (Marshall, Trans. 37); in small veins in serpentine in the Mikonui district (Bull. 6).

Magnetite (Magnetic Iron Oxide): Common rock-constituent. Fine octahedrons associated with the magnesian rocks of Westland (Bull. 1 and 6) and with chlorite schists of Otago. Very plentiful in the ironsands of the sea-beaches of New Zealand.
See also Ironsand.

Malachite (*Basic Cupric Carbonate*): North Westland (Bull. 1); Parapara (Bull. 3); Coromandel (Bull. 4); Whangaroa district (Bull. 8); Mount Radiant (Bull. 11); Thames (Bull. 10); Dun Mountain district (Bull. 12); and in almost all the localities where copper-ores occur.
See also Chalcopyrite.

Manganese-ores: The following list includes a few of the occurrences reported in the Lab. reports; the various minerals have not as a rule been specifically determined: Ashburton; Avoca, North Auckland; Bay of Islands; Berhampore, Kilbirnie, Makara, and Maranui, Wellington; Blenheim; Coromandel; Dome Mountain, Southland; Egmont National Park; Hikurangi, North Auckland; Hokianga; Kawau Island; Mangapai, Auckland; Mangonui; Napier; Nelson; Nenthorn, Otago; Ohario; Okaihau, Bay of Islands; Pahaki, Auckland; Paraparaumu; Parua Bay; Picton Sound; Queen Charlotte Sound; Rapaka, Bay of Islands; Red Island; Ruapekapeka; Russell; Stewart Island; Taieri Mouth; Te Awaiti, East Coast; Terawhiti; Thames; Mount Vernon, Napier; Waihi Mine; Waikato; Waimarama; Waimate, North Auckland; Whangarei; Whangaroa.
See also Description of Exhibits, p. 22.

Manganite (*Hydrated Manganese Sesquioxide*): Cavities in quartz, Tararu Creek (Skey, Rep. G.S. 1871). Cox (Trans. 14) reports the mineral from—Waiheke Island; Kawau Island; Bay of Islands; in gravels of Kawarau and Clutha Rivers, Otago; Whangarei; Waimarama; Tory Channel; Wellington. Also Waipu (Hector, Rep. G.S. 1871).

Marble (*Carbonate of Lime*): Caswell Sound; Parapara; Takaka-Motueka district; Maniototo; Lake Wakatipu; Eketahuna district; Whangaroa (Bull. 8); Crooked Arm, a branch of Doubtful Sound (Liversidge, Trans. 10); Kakahu, Canterbury (Monro, 1866).

Marcasite (White Iron-pyrites—*Iron Disulphide*): Whangaroa district, in various rocks (Bull. 8); Mount Radiant (Bull. 11); Waihi district (Bull. 15).

Margarite: Milford Sound (Hector, Cox, Trans. 15); West Coast district (Hector and Haast 1865); Rimu and Kanieri, near Hokitika, associated with Ruby (which see) (Bull. 6).

Meerschaum: Dun Mountain, Nelson (Skey, Lab. 6).

Melaconite (*Black Oxide of Copper, var. of Tenorite*): See Tenorite.

Melanterite (Copperas—*Hydrated Sulphate of Copper*): Stalactites in mines at Thames (Bull. 10) and Waihi (Bull. 15); Mount Radiant (Bull. 11); old drives in Coromandel district (Bull. 4).

Melilite: Plates in basanite at Puketeraki, near Dunedin (Marshall 1906).

Mellite: Thames (Skey, Lab. 6, and Hutton 1870); cave at Bligh Sound (Hector 1863).

Menaccanite: See Ilmenite.

Menilite (*var. of Opal*): See Opal.

Mercury (*Native*): Pakaraka, Bay of Islands (Skey, Lab. 5, 6, and 11); Waimate, North Auckland (Skey, Lab. 10); Whangaroa Survey District (Maclaurin, Lab. 42); Ohaeawai (Hutton, Trans. 3); also reported in amalgam with gold from Waipori (Cox, Trans. 14) and from Westport district (see Hector, Rep. G.S. 1892), but it is probable that the metal in these cases has been derived from sluicing claims.

Mercury-ore: See Cinnabar and Mercury (Native).

Mesotype: Banks Peninsula (Haast). Probably a variety of Natrolite, which see.

Metacinnabarite (?): Ohaeawai (Hutton, quoted in Bull. 8, p. 87).

Methane: See Gas—Natural.

Mica: See Muscovite, Biotite, Sericite, Fuchsite, Chrome Mica, Lepidolite, Lepidomelane, Paragonite, and Phlogopite.

Microcline (*one of the Feldspars*): Common constituent of many granitic and gneissic rocks and of a few other allied types of igneous rock. Particularly common in the granitic rocks of Stewart Island and of the Westport district.

Microperthite (*one of the Feldspars*): Occurs in many different varieties of igneous rock throughout New Zealand.

Mineral Waters: See Waters—Mineral.

Mirabilite (Glauber Salts—*Hydrous Sodium Sulphate*): In old drives at Coromandel (Bull. 4); Brancepeth, Wellington (Park 1910).

Mispickel: See Arsenopyrite.

Molybdenite (*Molybdenum Disulphide*): Richmond Hill, Parapara (Bull. 3); Bravo Island, near Stewart Island (Skey, Lab. 25); Mount Radiant (Bull. 11); Tararu Creek, Thames (Allen, Mines Record, vol. ii, p. 475): Wilson's Claim near Fourteen-mile Creek, Greymouth (Maclaurin, Lab. 44); Paparoa Ranges, Greymouth (Skey, Lab. 31); Dusky Sound (see Cox, Trans. 15).

Molybdite (*Molybdenum Trioxide*): Mount Radiant (Bull. 11); Fourteen-mile Creek, north of Greymouth (Morgan, MS.), encrusting molybdenite.

Monazite (*Phosphate of the Cerium Metals*): Montgomery's Terrace Sluicing Claim, near Blackball (Bull. 13); Bradshaw's Terrace, Westport (S. Fry, Mines Reports 1904 and 1905, and Maclaurin, Lab. 40); Collingwood district (Maclaurin, Lab. 46).

See also Rare Earths.

Mosandrite: In trachyte at Campbell Island (Marshall, Trans. 41).

Mountain Leather (*var. of Asbestos*): North Westland, in the Pounamu formation (Bull. 1); Paemoka, Auckland (Skey, Lab. 33).

Muscovite (*White Mica*): Common rock-constituent. Large plates at Dusky Sound (discovered by Docherty) and at Charleston, Westport district, in a pegmatite dyke, where it was worked for a short time. Analysis by Liversidge in Trans. 10, p. 498.

Nagyagite (*Sulpho-telluride of Lead and Gold*): Sylvia Mine, Thames (Park 1910).

Natrolite (*one of the Zeolites*): Needle-shaped crystals in basalts at Dunedin (Hector 1865); Oamaru; Whakahara (Cox, Trans. 15); Awatere Valley (Thomson, 1913); Banks Peninsula (Haast).

Natural Gas: See Gas—Natural.

Nepheline: Constituent of alkaline igneous rocks. Plentiful around Dunedin (Ulrich and others); Auckland, in basanite (Marshall, Trans. 40); boulders in New River gravels, North Westland (J. P. Smith, Trans. 40).

Nephrite (*var. of Amphibole*—Jade—*True New Zealand Greenstone*): Griffin Range and elsewhere in the Pounamu formation in North Westland (Bull. 1 and 6); in gravels in the Greymouth district (Bull. 13); Milford Sound (See Cox, Trans. 15); reported also at Cobb Valley, Nelson, and at D'Urville Island. Nephrite was known to the Maoris as pounamu, and was highly prized by them, owing to its use in making battle-axes, clubs (*meres*), and ornaments of various kinds.

See also Pounamu, and Description of Exhibits, pp. 25–26.

Nickel: Traces in copper-ore at Mount Radiant (Bull. 11) and Dun Mountain district (Bull. 12); Waihi, as oxide (Morgan); Karangahake. Pond reports the metal from Coromandel and from the following localities near and north of Auckland: Mahurangi, Manukau North Head, Papakura Valley, and Waipu. Additional occurrences (various Lab. reports) are—Cascade River, West Otago; Dusky Sound; Jackson's Bay district; Dun Mountain district; New Plymouth; Port Albert Road, Auckland; Richmond Hill, Parapara; South Westland; Whangarei.

See also Genthite, Gersdorffite, and Pimelite.

Nickel Glance: See Gersdorffite.

Nickel Silicate: (Probable) Mahurangi (Cox, Trans. 14). See also Genthite.

Obsidian or **Volcanic Glass :** Mayor Island, &c. (Hector). A rock : not recognized as a mineral species.
See also Tachylite.

Oil : See Petroleum.

Oil-shale : Occurs in a number of localities, the most important of which is Orepuki. Other occurrences are at Cambrian, Roxburgh, and near Dunedin.

Oligoclase (*one of the Feldspars*) : Common in all the more acid igneous rocks.

Olivine : Common in all the basic igneous rocks of New Zealand, and particularly so in peridotites such as the dunite of Nelson district and north-west Otago, which is almost wholly composed of olivine.

Onyx : See under Quartz.

Opal (*Silica with a varying amount of water*)—
Common Opal : Whangaroa Harbour (Bull. 8) ; cavities in rhyolite of Thames district (Bull. 10) ; in this and other allied volcanic rocks elsewhere in the Thames–Tairua–Waihi district (Bull. 15) ; Cabbage Bay, Coromandel ; Malvern Hills and Mount Somers (Haast).
Menilite : Bay of Islands (Hector, Rep. G.S. 1892).
Opal-Jasper : Portobello, Otago, in trachytic tufa (Liversidge).
Pitch Opal : Dunstan, Rakaia Gorge, and Harper's Hill (Liversidge).
Wood Opal or *Silicified Wood :* In many places in the Coromandel and Thames districts ; Mount Somers ; Waikaia, south-east Otago, where there is a fossil forest exposed at the sea-beach ; many other localities.
Precious Opal (*Fire-or Gem Opal*) *:* Cabbage Bay, Coromandel (Marshall 1912) ; Tairua, Hauraki Peninsula (Hector, Rep. G.S. 1892 and Bull. 15) ; Otago Peninsula (see Hector, Rep. G.S. 1892) ; also reported from Rangitata district, Canterbury, where, however, the opals proved to be Queensland gems. At Tairua the opal is of good quality but difficult to extract from the matrix.
See also Hyalite, Tripolite, Siliceous Sinter, and Geyserite.

Orpiment (*Arsenic Trisulphide*) : Coromandel (Maclaren 1900).

Orthite : See Allanite.

Orthoclase (*one of the Feldspars*) : Common as an important constituent of granites and other igneous rocks of New Zealand. Sanidine, a clear glassy variety, is very frequent near Dunedin in trachytes, bostonites, trachydolerites, &c. (Marshall 1906 and others).
See also Valencianite.

Osmiridium (*alloy of Osmium and Iridium*) : Arthur Creek, North Westland (Maclaurin, Lab. 42, but the locality is doubtful) ; Parapara (Skey, Lab. 30 and 31) ; Orepuki ; Waikaka, Otago.

Osmium : Parapara (Skey, Lab. 30 and 31).
See also Osmiridium.

Ouvarovite : See Uvarovite.

Ozocerite (Ozokerit—Mineral Wax—*a Simple Hydrocarbon*) : Brown coals of Dunstan (Hacket 1865). In several localities in eastern Wellington and the Gisborne – East Cape district a wax or spermaceti-like substance has been found embedded in the soil. It is regarded as an indication of petroleum.
See also Dopplerite.

Palagonite : Harper Hills, Canterbury (Haast 1865) ; Taipo Hills, Otago (Cox, Trans. 15) ; Two Brothers (Hector, Rep. G.S. 1892). Not now recognized as a mineral species.

Palladium : Parapara (Skey, Lab. 30, p. 16), associated with osmium, iridium, platinum, &c.

Paragonite (*Sodium Mica*) : Jacob's River, Westland (Marshall, Trans. 41).

Peacock Copper-ore : See Bornite.

Pearl Spar : See Dolomite.

Penninite (*a var. of Chlorite*) : Lake Wakatipu district (Skey, Lab. 16).

Perovskite (Perofskite—*Calcium Titanate*) : In Bell Hill diorite, Dunedin (Marshall 1906) ; Campbell Island (Marshall, Trans. 41).

Perthite (*one of the Feldspars*) : Constituent of various igneous rocks. Stewart Island and Dunedin (Marshall, Trans. 39 and Q.J.G.S. vol. lii, 1906) ; Westport district (G.S. 1913) ; New River gravels, North Westland (Smith, Trans. 40).

See also Microperthite.

Petroleum : Ohaeawai, North Auckland, associated with hot springs (Bull. 8) ; indications at Mount Radiant (Bull. 11) ; Kotuku and Dobson, near Greymouth (Bull. 13) ; Taranaki (Bull. 14) ; Gisborne district (Bull. 9) ; exudations from sandstone in Liverpool State Coal-mine, near Greymouth (Morgan 1913) ; Cheviot (Skey, Lab. 32 and 33) ; Waiotapu, near Rotorua (Skey, Lab. 26) ; Waiapu, East Cape district (Skey, Lab. 2) ; Weber district (traces). The only localities where important amounts of petroleum have been found are Taranaki, Kotuku, and Gisbórne districts.

See Description of Exhibits, pp. 20–22.

Petzite (*Gold-Silver Telluride*) : Sylvia Mine, Tararu Creek, Thames (Park 1910) ; (probable) at the Maratoto mines (Bull. 15).

Phagopite : Mentioned in Park's "Geology of New Zealand," 1910. Probably misprint for Phlogopite (which see).

Phlogopite (*Magnesium Mica*) : (?) Quartz-schist, Dunstan Range. See Phagopite.

Phosphates : Phosphate rock has been recorded from Wangapeka (Skey 1885) ; Weka Pass (concretions) (Skey 1887) ; also, in various Lab. reports, at Amberley ; Hoteo River, Kaipara ; Mangaharuru Survey District, Hawke's Bay ; Onewhero, Waikato ; near Whangarei. At Clarendon and Milburn in Otago large deposits of phosphate rock exist and are worked on a considerable scale. (See Description of Exhibits, pp. 23–4.) A phosphate-deposit in a cave at Wellington Heads has also been recorded (Maclaurin, Lab. 42). At Onewhero a limestone contains 10·60 per cent. of calcium phosphate (Maclaurin, Lab. 43). Aluminium phosphate occurs on Green Island, off the coast near Dunedin (Aston). Calcium phosphate is recorded from Campbell Island (Lab. 37, p. 11).

See also Coprolites ; Apatite.

Picotite (Chrome Spinel) : In harzburgite, Dun Mountain district (Bull. 12) ; South-west Otago (?) (Marshall, Trans. 41) ; in basalt, Leith Valley, Dunedin (Marshall 1906) ; Kakanui R. (Thomson, Geol. Mag. 1907).

Picrolite (*var. of Serpentine*) : Dun Mountain district (Cox, Trans. 15).

Picrosmine (*sectile var. of Serpentine*) : Dun Mountain (Cox, Trans. 15).

Pimelite (*a Nickel-bearing Silicate*): Malvern Hills and Clent Hills, Canterbury (Haast 1865).

Pinite: Mount Radiant (Bull. 11), Golden Ridge, Collingwood (Marshall, Trans. 41, p. 101), and elsewhere, as decomposition product in various rocks. Has no distinct mineral character.

Pistacite: See Epidote.

Plasma (*var. of Quartz*): See under Quartz.

Platinum: Takaka (Hochstetter); Stewart Island (Hector, Trans. 2); Thames, in reef in Queen of Beauty shaft (Pond, Trans. 15); Collingwood (Hochstetter); Orepuki and Round Hill, Southland; Clutha River gravels (Gordon, Miners' Guide, 1906); Wade (Pond—see Farquharson, Trans. 43); Gorge River (see Farquharson); north of Ten-mile Creek, Greymouth, in gravels (?); Taipo Gorge and Harley's Creek, North Westland (Bull. 1), in quartz veins; in pyritic vein, Serpentine Creek, Mikonui district, Westland (Bull. 6); Mangles Valley, Upper Buller (Skey, Lab. 7); Parapara (Bull. 3); near Westport at Christmas Terrace, Bradshaw's Terrace, Wareatea Creek, and Mount Rochfort Sluicing Claim (Maclaurin, Lab. 40, and Fry, Mines Record, vol. x, pp. 12–13); Owharoa, near Waihi (Haeusler, Trans. 23); Waikaia and Waikaka dredges, Otago (*R*); Worksop dredge, Maimai, Reefton (*R*). Except where specially indicated to the contrary, all the platinum is found in alluvial deposits.

Platiniridium: Takaka (Hochstetter); Mangles Valley, Upper Buller (Skey, Lab. 7); Queen of Beauty shaft, Thames (Pond, Trans. 15); Orepuki (Hochstetter).

Pleonaste (*Iron-Magnesia Spinel*): Kakanui R., in olivine nodules in calcareous tuffs (Thomson, Geol. Mag. 1907).

Pounamu: The Maori name for Nephrite (which see). Varieties of pounamu are Kahurangi, Inanga, Kawakawa, Auhunga, Totoweka, Raukaraka, and others. Kokotangiwai or Tangiwai (which see) is a variety of serpentine.

Prase (*var. of Quartz*): See under Quartz.

Prehnite: Usually a secondary mineral in cavities of basic eruptive rocks. Dun Mountain district, replacing grossularite in rodingite (Marshall, Bull. 12); Moeraki and Otepopo (Hector—see Rep. G.S. 1892); Snowy Peak Range (Daintree).

Proustite (Ruby Silver Ore—*Sulpharsenite of Silver*): Thames (Hutton); Waihi mines (Bull. 15); Westport district (Fry, Mines Record, vol. x, pp. 12–13).

Psilomelane (*Hydrous Manganese Manganate*): Widely distributed in small veins in the Whangaroa district (Bull. 8); Thames, Waihi, Waipu (Park 1910); with other manganese ores at Taieri Mouth and elsewhere (Marshall 1912); Seaview, Dunedin (Marshall 1906); Kawau, Waimakariri, Bay of Islands, Waiheke, and Wellington (Hector, Rep. G.S. 1892).

Pyrargyrite (Ruby Silver Ore—*Sulphantimonite of Silver*): Coromandel (Maclaren 1900); frequent in some of the Thames mines—*e.g.*, Kuranui-Caledonian (Bull. 10); Waihi mines (Bull. 15); Puhipuhi, near Whangarei (Park 1910); Great Barrier Island (*R*).

Pyrite (*Iron Disulphide*): Very common throughout New Zealand in quartz reefs and in various rocks. Some widespread rocks contain large amounts of this mineral.

Pyrolusite (*Manganese Dioxide, commonly with a little water*): In veins of Coromandel (Bull. 4), Thames (Bull. 10) and Waihi-Maratoto-Komata-Waitekauri (Bull. 15) goldfields; Waiheke and Kawau, with manganite (Park 1910).

Pyromorphite (*Lead Phosphate with a little Lead Chloride*): Tui Mine, Te Aroha, and Grace Darling Mine, Waitekauri (Park 1910).

Pyrope (*Magnesium-Aluminium Garnet*): Kakanui (Hector, Trans. 16, p. 541).

Pyroxene: See Ægirine, Augite, Enstatite, Hypersthene, Bronzite, Diallage, Diopside, &c.

Pyrrhotite (Pyrrhotine—*a Sulphide of Iron*): Dun Mountain district (Bull. 12); Richmond Hill, Parapara, and Dusky Sound (Skey, Lab. 13); Westland (Cox); Poerua (G. S. Ann. Rep. 1913); George Sound (Don); Mount Rangitoto, Westland (Don).

See also Troilite.

Quartz (*Silica*): Everywhere. Forms matrix of lode-gold, &c.

Rock Crystal: Milford Sound, Tamata, Keruru (Napier), Taupo district, Mount Somers, and Clent Hills, &c. (Cox, Trans. 15); various localities on the west side of the Southern Alps and elsewhere.

Rose Quartz: Rakaia Gorge (Haast).

Plasma: Clent Hills, Canterbury (Haast).

Prase: Clent Hills and Malvern Hills (Haast).

Agate: Malvern Hills (Haast); Herbertville (Skey, Lab. 23).

Amethyst: Rakaia Gorge (Haast); Tokatea Hill, Coromandel; Waihi Mine (Morgan, and Bull. 15); Duffer's Creek, near Lake Ianthe, Westland (Morgan in MS.).

Bloodstone: Clent Hills, Snowy Peak, and Malvern Hills, Canterbury (Haast); Herbertville (Skey, Lab. 19).

Carnelian: Coromandel district (Bull. 4); Malvern Hills; Mount Charles, Dunedin (Hector).

Chalcedony: Clent Hills, Gawler Downs, and Mount Somers, Canterbury (Haast); Tokatoka (Wairoa River), Moeraki, and Otepopo (Hector, Rep. G.S. 1892); Coromandel, in small fissures in argillites, and in cavities in the volcanic rocks there and in the Thames district (Bull. 4 and 10).

Onyx: Malvern Hills, Canterbury (Haast).

Chrysoprase: Moeraki, Otepopo, Dunedin, Coromandel, Canterbury, filling cavities in amygdaloidal rocks (see Hector, Rep. G.S. 1892).

Flint: Kaipara, Clarence Valley, and Campbell Island, in chalk-marls (Hector, Rep. G.S. 1892); Amuri Bluff, in limestone; Bay of Islands and Tapanui (Liversidge); Whangarei, in diatomaceous earth (Hector).

Jasper: Nests and veins in argillites and volcanic rocks of the Hauraki goldfields (Bull. 4, 10, and 15); Auckland, in tuffs and conglomerates (Hochstetter); Snowy Range, Canterbury (Haast); common in Otago in several localities; widespread in the so-called "jasper slates" of the "Maitai" rocks of the old Geological Survey.

Potato Stone or *Geode:* Snowy Ranges (Haast).

Agate-Jasper: Coromandel (Hector).

Lydian Stone: Gawler Downs, Canterbury (Haast); Whangarei (Cox, Trans. 15).

See also Siliceous Sinter, Silica, and Tridymite.

Rare Earths : Various occurrences of material containing Ceria, Thoria, &c., have been reported—*e.g.*, Bradshaw's Terrace, Westport (Maclaurin, Lab. 40); Montgomery's Terrace S.C., near Blackball (Bull. 13); Eric Creek, Nelson (Maclaurin, Lab. 43); and Tawhai, Reefton (Maclaurin, Lab. 40). None of these is of economic value.

See also Monazite.

Realgar (*Arsenic Monosulphide*): Coromandel (Maclaren 1900); crusts on ore in the Thames district (Bull. 10); crusts on sinter in the Whangaroa district (Bull. 8).

Resin (*Mineral*) (Retinite, Ambrite, &c.): In coal at Hyde, Caversham, Tuapeka, Waitahuna, Dunstan, Bay of Islands (Hector, Rep. G.S. 1892); in coal near Invercargill (Skey, Lab. 30), and at Blackball, Point Elizabeth, and other localities near Greymouth; also in coal at Seddonville, Charleston, Lower Buller Gorge, Pensini Creek (near Lyell), Orepuki, Waihao, Waikato, Whangarei, Kawakawa, &c. Occurs in sandstone or allied rock at Caversham, Waihao, and Nine-mile Bluff, Greymouth.

See also Kauri-gum.

Retinite : See Resin—(Mineral).

Rhodium : Parapara (Skey, Lab. 30).

Rhodochrosite (Dialogite — *Manganese Protocarbonate*): Tararu Creek, Thames (Skey, Lab. 6); Makara Valley, Wellington (Skey 1870); Waitekauri (Park); veinstone at Waihi and Waihi Beach mines (Bull. 15); with vein quartz at Coromandel (Bull. 4) and Thames (Bull. 10); Paraparaumu (Manawatu Railway-line) (Skey, Lab. 33 and 24). This last occurrence may be of some economic importance. See report by A. McKay in Mines Reports, 1899 (C.–9, pp. 2–3).

Rhodonite (*Manganese Metasilicate*): Coromandel, colouring vein quartz (Bull. 4); veinstone, Thames (Bull. 10); Waiheke (Skey); Kawarau, Dunstan, and Clutha (see Cox, Trans. 14); loose boulders Hokitika River, &c. (Bull 6); Queenstown district (Bull. 7); Hindon, Otago (*R*).

Richmondite (*var. of Tetrahedrite*): Richmond Hill, Parapara (Bull. 3 and earlier references—*e.g.*, Skey, Lab. 12).

Riebeckite (*var. of Amphibole*): In granite boulders near Lake Brunner and other parts of Westland (Smith, Trans. 40 and Bull. 13); Campbell Island (Marshall 1912).

Rock Crystal (*var. of Quartz*): See under Quartz.

Rubellane (? Rubellan—*an altered Biotite*): Banks Peninsula (Haast 1865).

Ruby — Oriental (*var. of Corundum*): North Westland (Bull. 1 and 6); occurs in boulders found in sluicing claims at Rimu and Kanieri near Hokitika. These have sometimes been referred to as "goodletite." They consist of a green micaceous material (probably a mixture of margarite and fuchsite) in which numerous rubies are embedded. Most of the rubies are not transparent enough to be called gem stones, but a few show promise of being valuable. (Bull. 6, pp. 123–124.)

Rutile (*Titanium Dioxide*): Mica-schists and gneiss of Central and West Otago, Dusky Sound and Cromwell (Park); granites and allied rocks near Westport (G.S. 1913); West Coast Sounds (Marshall, Trans. 39); Westland district, as rock-constituent (Bull. 6); beach sands at Orepuki.

Sanidine : See Orthoclase.

Sapphire (*var. of Corundum*) : Collingwood (Skey, Lab. 7).

Saussurite (*alteration product of Feldspar, not a mineral*) : Mount Torlesse, Canterbury (Haast 1865). Saussuritic material is common in altered feldspathic rocks. Hutton's saussurite-gabbro (Dun Mountain) has been shown by Marshall to be a grossularite-diallage rock, now known as rodingite.

See also Grossularite.

Scapolite : Dun Mountain, Nelson (Skey, Lab. 6).

Scheelite (*Calcium Tungstate*) : Pelorus Sound (Park 1910) ; Mount Radiant (*R*) (Bull. 11) ; Bendigo and Carrick lodes, Otago (Bull. 5) ; Queenstown district (Bull. 7) ; Waipori, Barewood, Buckleburn, Rees River, Richardson Mountains, Saddle Hill, and Mount Highlay, in Otago ; additional localities noted in Lab. reports are—Alexandra, Glenorchy, Lammermoor Range, Macrae's Flat and Tapanui Valley, in Otago ; Havelock, Armchair Creek, Mount Patriarch, Pine Valley, Jubilee Mine (Top Valley), and Wakamarina, in Marlborough.

See also Description of Exhibits, pp. 13–14.

Schiller Spar : See Bastite.

Schrötterite : (?) Malvern Hills (Liversidge, Trans. 10).

Selenium : In ore at Waihi and Maratoto (Alex. Montgomery, Mines Rep., 1889, C.–2, p. 20) ; Talisman Mine, Karangahake (Aston).

See also Selen-sulphur.

Selenite : See Gypsum.

Selen-sulphur : White Island (Liversidge, Trans. 10).

Senarmontite (*Antimony Trioxide*) : Waikari, Bay of Islands (Park), and Mount Radiant (Marshall—see Park 1910).

Sepiolite : See Meerschaum.

Sericite (*var. of Muscovite*) : Very common in most altered rocks.

Serpentine : Pounamu formation, North Westland (Bull. 1 and 6) ; Queenstown district (Bull. 7) ; Cromwell district (Bull. 5) ; Dun Mountain and "mineral belt," Nelson (Bull. 12) ; Red Hill, Caples River, Lake Harris, Mount Allen (Park 1910) ; Wade, near Auckland ; Milford Sound.

See also Iddingsite, Picrolite, Bowenite (Tangiwai), Asbestos, Chrysotile, Antigorite, Picrosmine, and Hectorite.

Siderite (Spathic Iron-ore—*Iron Protocarbonate*) : Seven-mile Creek, Nine-mile Creek, and south end of Paparoa Range, near Greymouth (Bull. 13) ; North Westland (Bull. 1) ; in schist, Otago (Hector) ; in the form of clay-ironstone with coal at many places—*e.g.*, Kakanui River, Mount Somers, Malvern Hills, Abbey Rocks (South Westland), and Raglan (Park 1910) ; Miranda Colliery, Manawatu Gorge, Baton River (Nelson), and Cairn Ranges (Canterbury) (Cox, Trans. 14) ; Whangaroa (Cox 1881).

"Silica" : A white powdery substance consisting almost entirely of silica is found in the gold veins of Thames (Bull. 10) and Waihi (Bull. 15).

Siliceous Sinter : This is deposited by the hot springs of Rotorua district, White Island, Taupo district, and elsewhere.

See also Geyserite.

Sillimanite : In schistose rocks in the Whangaroa district (Bull. 8) ; Stewart Island and Dusky Sound (Marshall, Trans. 41) ; Resolution Island Marshall, Trans. 39).

Silver—*Native :* Karangahake mines (G.S. 1912) ; Waihi Mine (Morgan) ; Thames and Omahu, in some of the mines (Bull. 10) ; gravels of the Shotover River, Otago (*R*) ; Kawau Island, Lake Wakatipu, Waipori (Hector, Rep. G.S. 1892) ; also alloyed with gold in electrum. Most of the gold from lodes contains more or less silver.

Smaragdite (*var. of Amphibole*) : Red Hill, Collingwood, in diorite (Hector, Rep. G.S. 1892); Kakanui River, North Otago (Thomson, Trans. 38).

Smectite (*var. of Halloysite*) : Great Barrier Island (Hutton).

Soapstone : See Talc.

Sodalite : Rock-constituent. Common in the alkaline rocks near Dunedin (Marshall and others) ; in ditroite found in gravels in North Westland (J. P. Smith, MS.).

Sphærosiderite (*var. of Siderite*) : In dyke rocks at Banks Peninsula and Mount Somers, Canterbury (Haast 1865 and Trans. 11).

Spathic Iron-ore : See Siderite.

Sphalerite (Zinc Blende—*Sulphide of Zinc*) : Bendigo, Otago ; near Dannevirke ; Caswell Sound (Skey, Lab. 17) ; King Country (Skey, Lab. 23) ; Parapara district (Bull. 3) ; mines at Coromandel (Bull. 4), Thames district (Bull. 10), and Waihi and Waitekauri districts (Bull. 15) ; Collingwood (Cox, Trans. 14) ; in Taranaki ironsands (Bull. 14) ; mines in Te Aroha district (G.S. 1912) ; Great Barrier Island (Hector, Hutton) ; Mount Radiant (Bull. 11) ; Whangaroa district (Bull. 8) ; Greymouth district (Bull. 13) near Roa ; Stewart Island (Skey, Lab. 24) ; Fitzherbert (Skey, Lab. 24) ; Red Hill (Skey, Lab. 23).

Sphene : See Titanite.

Spinel : Waipori (Hector 1865) ; Manawatu, Mount Somers (Canterbury), (Hector, Rep. G.S. 1892) ; Stewart Island (Skey, Lab. 24) ; Pahiatua (Skey—see Park 1910).

See also Hercynite, Gahnite, Pleonaste, and Picotite.

Staurolite : Golden Ridge, Collingwood district (Marshall, Trans. 41, p. 101).

Steatite : See Talc.

Stibnite (Antimonite—*Antimony Trisulphide*) : Coromandel (Bull. 4) and Thames (Bull. 10) in gold veins ; Alexandra, Carrick Range, Barewood, Waipori, Hindon, and elsewhere in Otago (Bull. 7 and other reports or papers) ; Reefton, in quartz lodes ; Endeavour Inlet ; Lake Ngawha, Whangaroa (Bull. 8) ; Langdon's Creek, Greymouth (Bull. 13 and other reports). Antimony-ores, chiefly stibnite, have also been recorded (various Lab. reports) from—Blenheim ; Buller district ; Collingwood ; Dannevirke ; Mount Edgecumbe ; Geraldine, South Canterbury ; Great Barrier Island ; Hokitika ; Kaikoura ; Kawarau Gorge, Otago ; Mahakipawa ; Milton, Otago ; Otaio, Canterbury ; Paparoa Ranges, Greymouth ; Picton ; Puhipuhi ; Rawhiti, Bay of Islands ; Resolution Bay ; Tararua Ranges ; Waimate, North Auckland ; Westport ; Whangapoua ; Whangarei ; Redman Creek, Ross.

See also Cervantite, &c.

Stilbite (*one of the Zeolites*) : Waihi district, in altered rhyolites and andesites (Morgan 1902 and Bull. 15) ; Tokatoka and Pukekorero, North Auckland (Cox, Trans. 15) ; Kauaeranga River, Thames (Bull. 10) ; Te Puke district (G.S. Ann. Rep. 1913) ; Dunedin (Liversidge, Trans. 10) ; Turnagain Point, Rangitata River, Canterbury (Haast 1865) ; Dun Mountain district, Nelson (*R*) ; Karori and Mangawai (Hector, Rep. G.S. 1892).

Stream Tin : See Cassiterite.

Sulphur : White Island ; Rotorua and Taupo districts, deposited from geysers, fumaroles, and hot springs ; Lake Ngawha, Whangaroa (Bull. 8) ; Waipara district, Canterbury, as efflorescence from sandstones (Haast 1870) ; Wangapeka, as efflorescence from pyritous reefs (Davis) ; near McKenzie, Cheviot, in connection with petroleum seepages (Morgan MS.) ; deposited by most hot springs, and also as an efflorescence in many coal and other mines where there is much pyrite present.

See also Description of Exhibits, p. 22.

Tachylyte : Described by Haast (Trans. 11) from Banks Peninsula; occurs also at Oamaru. Not now regarded as a mineral.

See also Obsidian.

Talc (Soapstone or Steatite) : Taipo Gorge, Griffin Range, Mount Jumbletop, Mount Bowen, and elsewhere in North Westland, associated with the Pounamu formation (Bull. 1 and 6) ; in talc-serpentine rock and talc-schists in the Mikonui district, Westland (Bull. 6) ; Milford Sound (Hector 1865) ; Martin's Bay (Skey, Lab. 21) ; Arrow district, Otago (Skey, Lab. 31) ; nests in serpentine at head of Springburn, Cromwell district (Bull. 5) ; in schists of Western Otago (Park 1910) ; West Coast Sounds (Hector 1865) ; Caples River, Lake Wakatipu (Marshall 1912) ; Jackson's Bay (Macfarlane, Rep. G.S. 1877) ; Collingwood (Hector, Bull. 3).

Tangiwai : Maori name for a nearly transparent green variety of serpentine, flecked with dark spots. It should not be confounded with the much harder pounamu. Found at Milford Sound, &c.

See Bowenite.

Taranakite : Sugarloaves, New Plymouth (Skey). This mineral was analysed and described by Skey as a new mineral. It approaches Wavellite. See Hector (Rep. G.S. 1892) for an analysis.

See also Phosphates.

Tellurium : In ore in the Coromandel mines (Allen, Mines Record, vol. iv, p. 469) ; Karangahake mines (Skey) ; several mines at Maratoto and at Rosemont Hill, Waihi (*R*) (Bull. 15).

Tenorite (*Cupric Oxide*) : Reported from Dun Mountain district copper-mines (See Bull. 12) and D'Urville Island (Hector). The black earthy variety, Melaconite, occurs also at Parapara (Bull. 3), Thames (Bull. 10) and Mount Radiant (Bull. 11).

Tetradymite (*Telluride and Sulphide of Bismuth*) : Probably present in the Waitangi and Monowai mines, Thames (Bull. 10).

Tetrahedrite (*Sulphantimonite of Copper*) : Coromandel (*R*) (Bull. 4) ; Richmond Hill, Parapara (Cox, Trans. 14 and Bull. 3) ; Koputaiaki Bay, Coromandel, in andesite associated with pyrites (Park 1910).

See also Richmondite.

Thorium : See Rare Earths and Monazite.

Titanite (Sphene—*Calcium Titano-silicate*) : A fairly common rock-constituent. Occurs sparingly in mica-schist of Central and West Otago and in granites of the West Coast. It is moderately abundant in dioritic rocks near Westport (G.S. 1913).

Titanium-phosphate : Antipodes Island (Aston).

Topaz : In granite at Stewart Island (Skey, Lab. 24) ; in gravels at Westport (Skey, Lab. 23), and at Chatto Creek, Arrow River, and Waipori (Hector, Rep. G.S. 1892).

Torbanite : Mangonui, North Auckland (Skey, Lab. 31) ; Miranda district Auckland (Skey, Lab. 26).

Tourmaline : In granite and gneiss on the West Coast (Hector and Haast) ; Resolution Island (Hector) ; Mosquito Hill, South Westland (Hector) ; Westland (see Bull, 1, 6, and 13) ; granite at Dusky Sound and Tata Island (Cox) ; Bedstead Gully, Collingwood (Hector) ; as large crystals in chlorite-schist at Collingwood (Park 1910) ; in hornfels and granite near Westport (G.S. 1913) ; Pensini Creek, near Lyell (G.S. 1913) ; Paterson's Inlet, &c.

Tremolite (*var. of Amphibole*) : In schists in North Westland (Bull. 1 and 6), particularly in the serpentine-dunite of Mount Bowen (Bull. 6) ; Lake Te Anau district (G.S. 1913) ; in quartzite in the Dunstan Range and in gneissic schist at Dusky Sound (Park 1910) ; Parapara (Cox, Trans. 15).

See also Nephrite.

Tridymite (*Silica*) : In volcanic rocks at Lyttelton (Ulrich—see Cox, Trans. 15 and Marshall, Trans. 26) ; Waihi (Rutley, Morgan, and others) ; Auckland, in tephrite (Marshall—see Park 1910).

Tripolite (*Silica with some water*) : The earthy variety—*Infusorial Earth*—is found at Cabbage Tree Swamp, Auckland (Maclaurin, Lab. 43) ; Mount Eden (Maclaurin, Lab. 39) ; New Plymouth (Maclaurin, Lab. 44).

See also Diatomaceous Earth.

Troilite (*Ferrous Sulphide*) : South Westland (Skey, Lab. 11) ; Champion Copper Mine, Nelson (Skey, Lab. 19). Probably a form of Pyrrhotite (which see).

Turgite (*intermediate between Hæmatite and Limonite*) : Parapara iron-ore (Bull. 3).

Uralite (*Pyroxene altered to Amphibole*) : Common in many igneous rocks —*e.g.*, West Otago, Westland, Nelson, the Bluff, and elsewhere.

Uvarovite (Ouvarovite—*Calcium-Chromium Garnet*) : Dusky Sound (Park 1910).

Valencianite (*Secondary Orthoclase*) : Common in the veinstone at the Waihi and Karangahake mines, where it was first identified by Lindgren.

Vanadium : In ironsand at Patea (See Bull. 14) ; in coal near Wharekirauponga, Waihi district (Morgan 1902).

Vesuvianite : See Idocrase.

Viridite (*a decomposition product of no definite composition*) : Gisborne district (Bull. 9) ; Smyth Creek, Mikonui district, Westland (Bull. 6). This aggregate occurs very generally, particularly in rocks of igneous origin.

Vivianite (*Hydrous Ferrous Phosphate*): Invercargill (Skey, Lab. 27); Urenui, Taranaki (Skey, Lab. 14); Wairarapa (Skey, Lab. 13); in bones at Dunedin and Awatere (Hector, Rep. G.S. 1892); Timaru, Pohangina, Port Chalmers, and Thames (Park 1910); near Waiuta, Reefton district, in quartz lode (Henderson and McPadden); Mercer; Riverton; Kingston.

Wad (*Hydrous Oxide of Manganese*): In quartz reefs of Thames (Bull. 10), Coromandel (Bull. 4), Waihi (Bull. 15), and Te Puke; Whangarei (Skey, Lab. 19); Stewart Island (Skey, Lab. 24); and many other localities.

Water, Artesian: There are many localities where artesian water may be obtained. The chief of these are the neighbourhood of Christchurch, Wanganui, and Napier.

Waters, Mineral: New Zealand is remarkably well supplied with mineral waters, both potable and therapeutic. No account of the various occurrences can here be attempted.

Wavellite: At the Sugarloaves, New Plymouth (Skey) associated with taranakite (which see). Aluminium phosphate occurs on Green Island, off the coast near Dunedin (Aston), and may perhaps be referred to Wavellite.

Witherite (*Barium Carbonate*): Thames (Skey, Rep. G.S. 1871, p. 85).

Wolframite (Wolfram): Stewart Island (Skey, Lab. 24); Westport district, in alluvium (Fry, Mines Record, vol. x, pp. 12–13); Mount Mantell (Maclaurin, Lab. 46).

Wollastonite (*Calcium Metasilicate*): Mikonui district, Westland (doubtful) (Bull. 6); Dun Mountain (Skey, Lab. 6); Westport district, in loose boulder (G.S. 1913).

Wood Opal: See Opal.

Wulfenite: (*Lead Molybdate*): Dun Mountain, Nelson (Cox, Trans. 14).

Zeolites: See Chabazite, Analcite, Natrolite, Heulandite, Stilbite, Gmelinite, Laumontite, &c.

Zinc (*Native*): There is a possible occurrence of this metal at Hape Creek, Thames (Park, Trans. 24, p. 384).

Zinc Blende: See Sphalerite.

Zincite (Red Oxide of Zinc—*Zinc Oxide*): Bedstead Gully, Collingwood (Skey—Park 1910).

Zinc Spinel: See Gahnite.

Zircon: In granites and gneissic rocks on the West Coast in microscopic crystals; Doubtful Inlet (Cox); Dusky Sound (Hector); in beach and river sands in Westland and Westport districts—*e.g.*, Whitcombe River, Waimangaroa River (Bull. 6; G.S. 1913); in porphyry, Campbell Island (Marshall 1912).

Zoisite (*one of the Epidote minerals*): In schistose rocks at Whangaroa; in gneissic rocks in North Westland and Karamea (Bull. 1, 6, and 11): Parapara district (Bull. 3).

NOTE.

All persons desiring to have mineral specimens identified may forward them to the Geological Survey Office, Wellington. Samples requiring chemical analysis should be sent to the Dominion Analyst. In all cases specimens should be carefully labelled, and the name and address of the sender enclosed. No charge for identification of minerals or for assay of prospectors' samples is made, provided the locality is stated.

By Authority: John Mackay, Government Printer, Wellington.
[800/12/13—20361